Bibliographic information published by the German National Library:

The German National Library lists this publication in the National Bibliography; detailed bibliographic data are available on the Internet at http://dnb.dnb.de .

Imprint:

Print and binding: Books on Demand GmbH, Norderstedt Germany
ISBN: 9783668708228

This book at GRIN:

https://www.grin.com/document/426006

Roland Nathan Kalonji

System Engineering and Energy programs

A Case Study on the University of the Witwatersrand in Johannesburg

GRIN Verlag

ENERGY PROGRAM

Case study: ***University of the Witwatersrand, JHB***

September 2017

Table of Contents

List of:

a.Figures

b.Tables

Abstract

First off, this project gives insight into how to define various systems management (project management and systems engineering) terms and to answer to the frequently Asked Questions in systems management. Specifically, to think like a systems engineer by changing the way of approaching and specifying a solution or addressing problems. Then the overall purpose is to use and apply Systems Engineering approaches to solve the high levels of unstructured problem situation and try to structure a debate concerning actions for improving the problem "energy program" situation within the System (University of Witwatersrand-Johannesburg).

1. Introduction

Case: *As a new systems engineering graduate the University of Witwatersrand has appointed you to drive the aggressive energy efficiency program that would see the university cut its energy bill by 50% in eighteen (18) months. Central to the statement of work (SoW) is the installation of solar panels on the roof of each building and introduction of smart energy efficient solutions for all energy consuming infrastructure, equipment, utilities and machinery.*

1.1 Stakeholders Analysis

The stakeholders are many, and working in complex interrelations on several programs. Then building and installation programs tend to focus on technological stakeholders; and in large programs the institutional stakeholders play a key role as well.

The section below presents the primary stakeholders involved in Solar Panels installation:

Institutional Stakeholders	Building Authority, Authority Having Jurisdiction, Energy company, promoter, General contractor, Subcontractors, the financier.
Technological Stakeholders	System Engineer, Program manager Team, Technical consultancy, architect, evaluator, Engineering, Procurement and Construction (EPC), Trainers.
Customers	Consumer , Clients

Table 1 Primary Stakeholders: "Author's own work"

Description

- *The Authority Having Jurisdiction will have varying construction and installation codes that must be complied with. This can encompass anything from building permits and fire and electrical code, to wetlands and endangered species protection.*
- *The Program manager team and the technical consultancy will be responsible for ensuring the viability of the program, site selection and acquisition, negotiating contracts, selecting contractors, and ensuring the program is built.*

- *The Engineering, Procurement and Construction (EPC) is the firm that will manage the installation process, procure the necessary materials, design and engineer the solar panels.*

2. TASK#1: Situation Analysis via - Peter Checkland's SSM

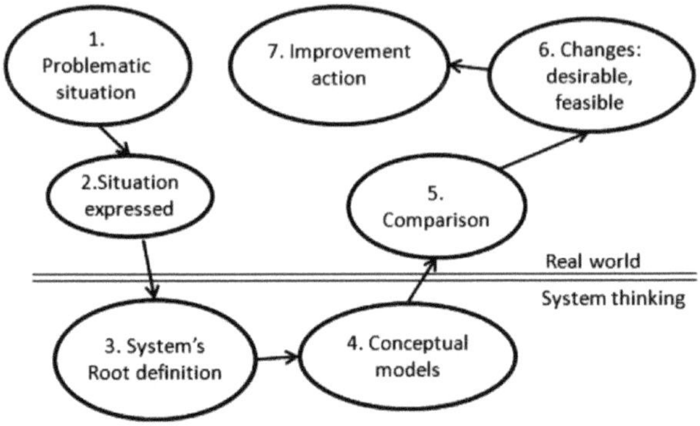

Figure 1 the SSM Model [1]

2.1 Step n○1: Problem Situation: Unstructured

2.1.1 The Naive Picture:

- No reliable information available about the current infrastructure, equipment, utilities and machinery, physical location and capacity of the University buildings.
- Power Purchase Agreements (PPAs), Energy Savings Performance Contracts (ESPCs), and Utility Energy Services Contracts (UESCs) are not known.
- Energy bill per month is not specified (amount of power KWh).
- Solar resource throughout Gauteng in kilowatt hours per meter squared per day (kWh/m2/day) not known.
- Lack of tools to help assess the technical and economic potential for the Solar PV at the University of Witwatersrand.

2.1.2 The Problem Statement

The University of Witwatersrand is having difficulty in fully optimizing its energy bill because of challenges involved in energy consuming infrastructure, equipment, utilities and machinery.

2.2 Step n∘2: Problem Situation: Expressed

2.2.1 The Rich Picture:

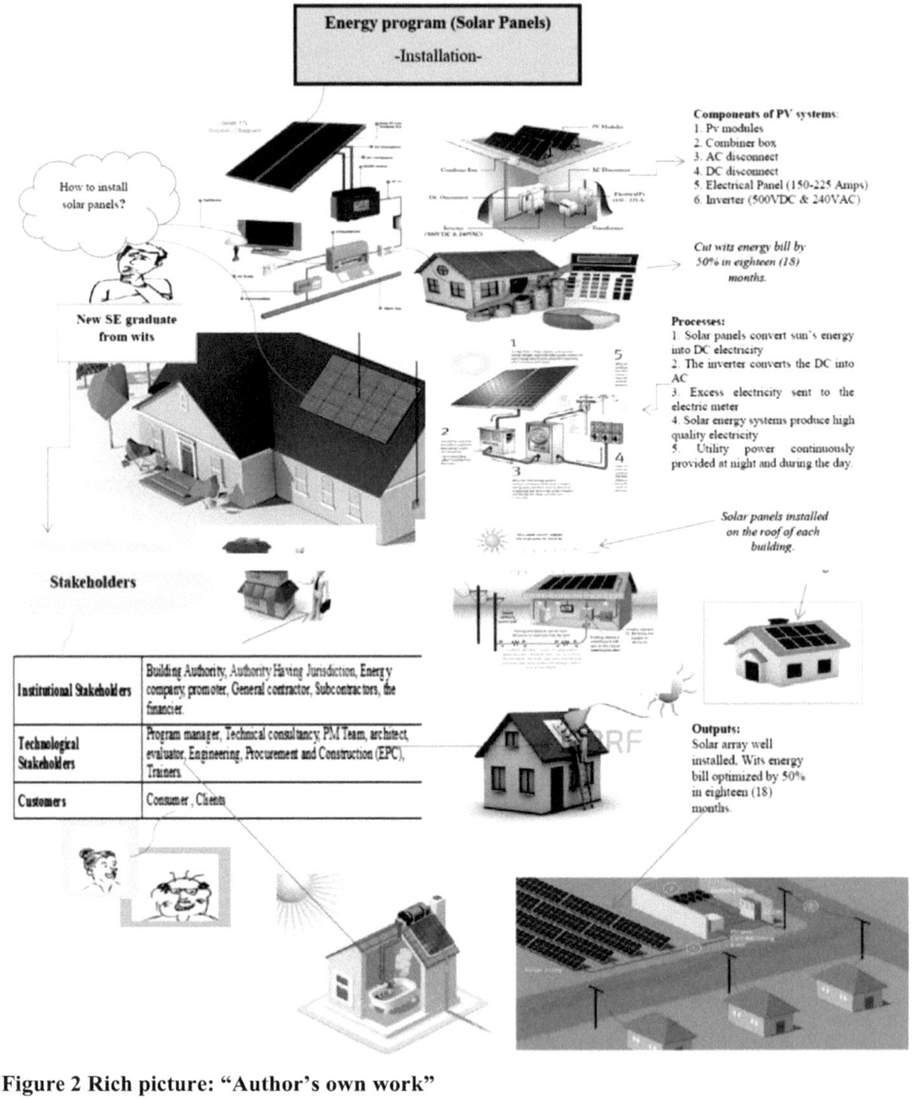

Figure 2 Rich picture: "Author's own work"

2.2.2 The Relevant Systems

The Ellies Renewable energy can be taken as relevant systems because they have always been on providing quality products and exceptional service to the customer; serve customers' needs in Communication, Lighting and Power.

2.3 Step n◦3: Root definition on relevant system

2.3.1 Root definition of the Ellies Renewable energy

A system to match consumer needs with the leading technologies to deliver holistic solutions and the power of the latest technology; the power to find an energy solution that won't cost you the earth [2].

2.3.2 The CAPETOWN MODEL

C-	**Customers**	Consumers
A-	**Actors**	Employees of Ellies, qualified electricians
P-	**Performance**	To find energy solutions that is both green and affordable.
E-	**Environment**	Terms and Conditions; legal set up.
T-	**Transformation**	To provide some of the world's leading power-saving technology.
O-	**Owners**	Ellies board of directors, executive managers, team of specialists.
W-	**Weltanschauung**	Committed to bring the latest in product by offering and placing the needs of the customers first always.
N-	**Nature of the System**	Human-made, open system.

Table 2 CAPETOWN Model (Ellies renewable energy): "Author's own work"

3. TASK#2: Outline:

3.1 Five functions of a Systems Engineer

The systems engineer functions involve:

- Technical Planning and strategy: the systems engineer collaborates with executive team members, decision makers, and stakeholders to define requirements and systems goals, and to identify and resolve systems issues.
- Operational Management: the systems engineer proposes and creates system design models, specifications, diagrams, and charts to provide direction to the development teams.
- Stability, integrity, and efficient operation within the systems life cycle.
- Organizing and staffing (chart working groups, reviews, risk Management, recruiting, training and team Building).
- Controlling and directing (Baseline Control, System Design Team Meetings, Requirements development, Verification and Validation, Performance Appraisals).

3.2 The purpose of the Systems Engineering Management Plan (SEMP)

The eventual purpose of the SEMP is to:

- Provide the framework and guidance for all engineering activities within the overall energy program.
- Give unique insight into the responsibilities and authorities of the active participants (specialist engineers, design engineers, test engineers, component design engineers, subcontractors, etc.)
- Outlines the Systems Hierarchy, the high-level WBS, major phases, major deliverables and technical objectives.

3.3 Systems Engineering Management Plan (SEMP)

3.3.1 Statement of Work (SoW)

Install solar panels on the university's building roof (to include the necessary batteries, transformers, and connections) to generate solar power electricity for the building's operations in order to reduce the amount of power that the University must purchase.

3.3.2 Scope of Work

This System Engineering Management Plan (SEMP) will describe the approach and phases that will be used during the program to support and eventually ensure the successful development, deployment and commissioning of the Ellies Solar Panels, including operations and maintenance of the system .

a. Technical Objectives:

- Make sure that equipment is installed according to the design with minimal changes and products are fully operational.
- Installation is completed on-time and within budget.
- Building maintenance staff receives training and is fully capable of operating and maintaining the system.

b. Assumptions and Constraints

Assumptions

- Permits are already acquired
- Normal daytime working hours allowed
- Existing site conditions on the University and around the building will not impede installation.

Constraints

- Budget for the installation and materials will be fixed once contracts are awarded
- Building will remain occupied and university operations will continue throughout the program; will not include relocation of any employees or operations during installation.
- Long lead time required for the procurement of some materials.

3.3.3 Ellies System Structure

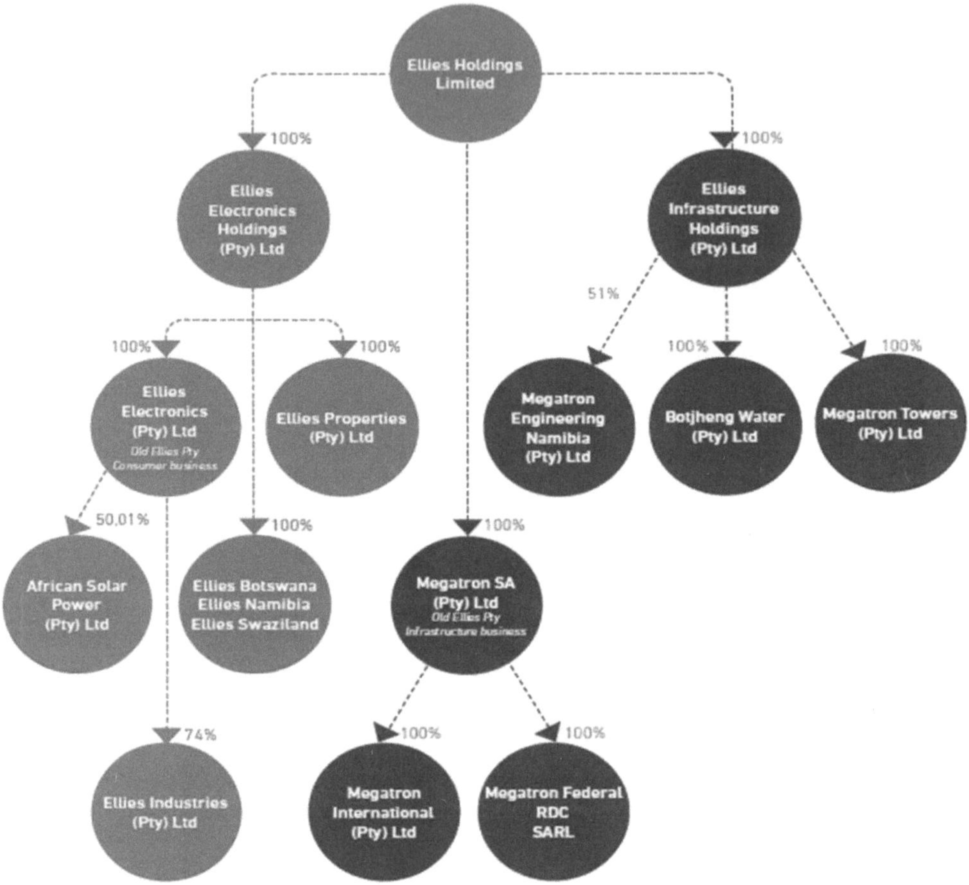

Figure 3 Ellies Corporate Structure [3]

3.3.4 Work Breakdown Structure (WBS)

Figure 4 Work Breakdown Structure (MindGenius 60002): "Author's own work"

a. Schedule Planning

Finally the program task needs to be scheduled systematically by considering durations and predecessors of each activity as shown in the figure below:

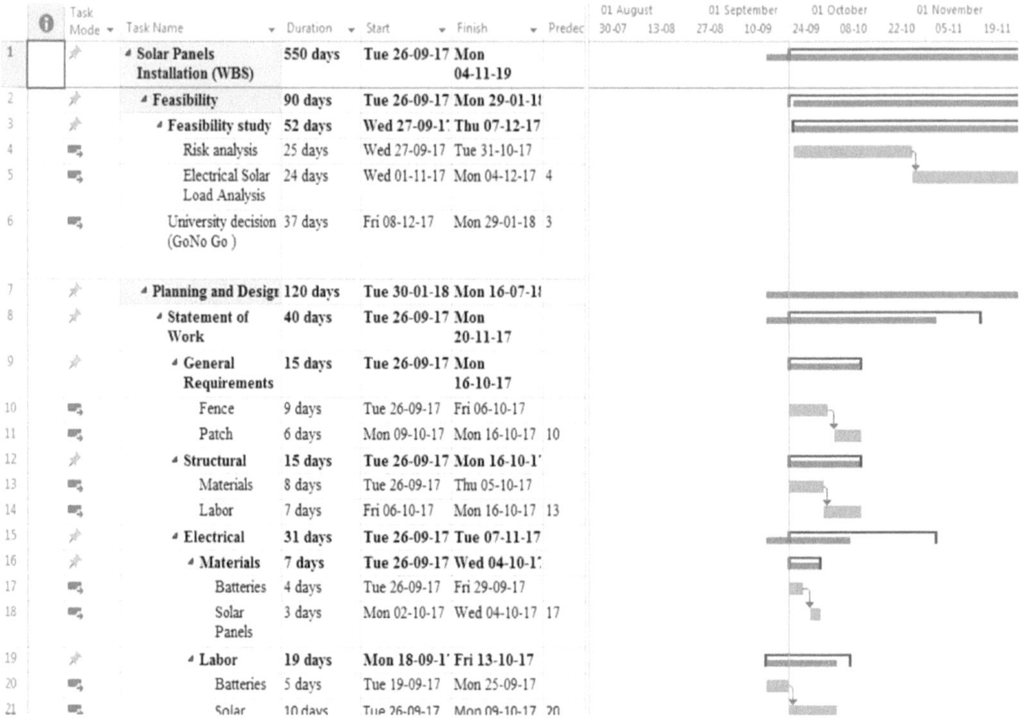

	Task Name	Duration	Start	Finish	Predec
1	**Solar Panels Installation (WBS)**	**550 days**	**Tue 26-09-17**	**Mon 04-11-19**	
2	**Feasibility**	**90 days**	**Tue 26-09-17**	**Mon 29-01-1!**	
3	**Feasibility study**	**52 days**	**Wed 27-09-1'**	**Thu 07-12-17**	
4	Risk analysis	25 days	Wed 27-09-17	Tue 31-10-17	
5	Electrical Solar Load Analysis	24 days	Wed 01-11-17	Mon 04-12-17	4
6	University decision (GoNo Go)	37 days	Fri 08-12-17	Mon 29-01-18	3
7	**Planning and Desigr**	**120 days**	**Tue 30-01-18**	**Mon 16-07-1!**	
8	**Statement of Work**	**40 days**	**Tue 26-09-17**	**Mon 20-11-17**	
9	**General Requirements**	**15 days**	**Tue 26-09-17**	**Mon 16-10-17**	
10	Fence	9 days	Tue 26-09-17	Fri 06-10-17	
11	Patch	6 days	Mon 09-10-17	Mon 16-10-17	10
12	**Structural**	**15 days**	**Tue 26-09-17**	**Mon 16-10-1'**	
13	Materials	8 days	Tue 26-09-17	Thu 05-10-17	
14	Labor	7 days	Fri 06-10-17	Mon 16-10-17	13
15	**Electrical**	**31 days**	**Tue 26-09-17**	**Tue 07-11-17**	
16	**Materials**	**7 days**	**Tue 26-09-17**	**Wed 04-10-1'**	
17	Batteries	4 days	Tue 26-09-17	Fri 29-09-17	
18	Solar Panels	3 days	Mon 02-10-17	Wed 04-10-17	17
19	**Labor**	**19 days**	**Mon 18-09-1'**	**Fri 13-10-17**	
20	Batteries	5 days	Tue 19-09-17	Mon 25-09-17	
21	Solar	10 days	Tue 26-09-17	Mon 09-10-17	20

Figure 5 Scheduling Planning (MS-Project): "Author's own work"

b. Major Deliverables

The eventual aim of the program is to deliver a cost effective 50% reduction in energy within 18 (eighteen) months. Then the following deliverables are in the scope of this program:

- Installation of solar panels, structural improvements, transformers, batteries, a battery shelter
- Facility, and necessary connections, wiring, and meters.
- Training of building maintenance team on operation and maintenance of the installed solar panels and associated equipment.
- Installation drawings and specifications.

- Generation of solar panels to power the University's building.
- Feasibility study to determine cost and, cost savings for the program.

c. Major phases

The program can be divided into 5 major phases:

- Feasibility
- Planning and Design
- Installation
- Training
- Turnover

d. Responsibility Assignments

WBS#	Deliverable Description	Resources					
		Owner	SE	PM Team	Techn. Cons.	Gen. Contractor	Trainer
1.1	Feasibility Study	S	R	I	I		
1.2	University decision	S	I				
2	Planning/Decision		I		I		
2.1	Statement of Work	R/S	I	I	I		
2.2	Work Schedule and Cost estimate	R	I	I			
2.3	Request for quote	S	I				
2.4	Write Contract	I	S	I		S	
3	Installation Solar		I	I	I	S	
4	Training		S				I
5	Turnover	R	I			I	
5.1	Review Acceptance Documents		S	I		I	
5.2	Accept.Docs To owner	S	I				

LEGEND: O – Owner   I – Input Required   R– Reviewer   S – Sign off Required

Table 3 Responsibility Assignments: "Author's own work"

3.3.5 Risk Management

ID	Risk Description	Affected Area/s	Probability Of Occurrence	Impact on Project	Priority	Strategies/ Actions (Avoidance, Transference, Mitigation, Acceptance)	Owner	Status (Open or Closed)
1	Unexpected Weather Patterns	Schedule/ Cost	Medium	High	1-High	Acceptance		Closed
2	General Contractor/ Subcontractor Solvency	Schedule/ Cost	Low	High	2-Medium	Transference	PM Team	Open
3	Excessive Lead Times for materials	Schedule	Medium	High	1-High	Transference	PM Team	Open
4	Injury to installation Worker	Schedule/ Cost	Medium	High	1-High	Transference/ Mitigation	PM/ Contractor	Open
5	Scope creep/ Mismanagement of Changes	Schedule/ Cost	Medium	High	1-High	Mitigation	PM	Open
6	Schedule Overrun	Schedule/ Cost	Medium	High	1-High	Mitigation	PM	Open
7	Budget Overrun	Cost	Medium	High	1-High	Acceptance		Closed
8	Design Problems	Scope Cost/ Quality/ Schedule	Medium	High	1-High	Mitigation	PM/ Techn.Cons.	Open

Table 4 Risk Management: "Author's own work"

3.3.6 Operation and Maintenance

There are several duties during the operation and maintenance of the Solar panels. The process is highly relying on the maturity of the technology. The duties are to keep the array clear from any trees, monitor the electronic system failures, mechanical tracking system maintenance and replacing and servicing the damage component inside the system [4].

4. Conclusion

In conclusion, this project gave insight into:

- **The techniques**: How to design the overall system architecture and the technical approach of a system; to know the generic model of Systems Engineering Life Cycle and solve the high levels of unstructured problem situation via the 7 Peter Checkland's SSM and the SEMP.
- **The application of systems management principles in case studies and projects:** Identify and deal with the most important issues in a system development process; understand the realities of complex system design, the stepwise development of complex systems, and the organization of system development projects.
- **To prioritize system requirements and satisfy stakeholders' expectations**: ensure that various system attributes are appropriately weighted when balancing multiple technical efforts and stakeholders' needs are met throughout a system life cycle.
- **To decide which risks are worth undertaking and which are not**: identify, assess, manage and monitor the risks to which the system is exposed.

5. References

[1] Checkland, P., & Scholes, J. (1999). Soft systems methodology: a 30-year retrospective (pp. 330-pages). Chichester: John Wiley.

[2] http://www.elliesrenewable.co.za/wp-content/uploads/2016/09/Off-Grid.pdf

[3] http://www.elliesrenewable.co.za/wp-content/uploads/2016/09/Off-Grid.pdf

[4] GPEKS Clean Energy Developments and Services (2010). Feasibility Analysis for 520kW Solar Photovoltaic Project at Consejo Belize /http: //solartechnologies. ru/

Appendices

Appendix A: Rich Picture (Hand sketched)

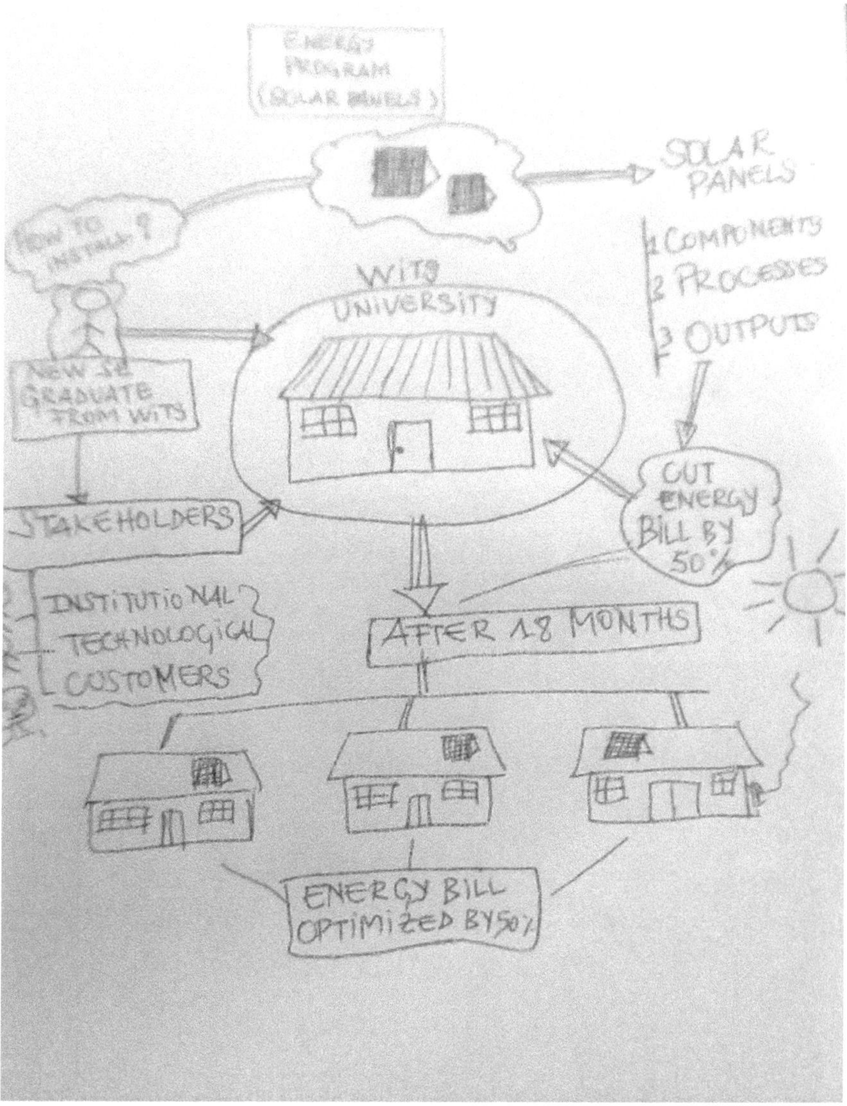

Figure 6 Rich Picture: "Author's own work"